Why the Earth is Getting Hotter
(A Brief Essay on Solar Warming)

By Curtis R. Crim, BA, CEO

ISBN: 978-0-9888255-1-2

Printed in the United States of America

First Printing

Dedication

Dedicated lovingly to Celeste Elaine Cool.

TABLE OF CONTENTS

Introduction:
The Theory

In this book, I am not disputing whether Global Warming is an actual event that we are experiencing on the Earth at this time.

The evidence is overwhelming, and if you think that Global Warming does not exist, then you are an idiot, and in denial.

I am not going to add any information pertaining to the popular working theories related to the cause of global warming, but rather am introducing an additional theory that is (probably) *part* of the cause of what we see happening to our global environment presently.

Although the consequences of what is happening to the Earth affect all of us, the effect commonly known as "Global Warming" is probably caused by more than one factor.

Chapter 1:
Global Warming

I thoroughly disapprove of how the corporations have destroyed the global environment, culture and political systems; however, they are probably not 100% responsible for the collapse of the Earth's environment.

Sure, it was a horrible idea to dump tons of crude oil into the oceans, but this alone only contributes to global warming.

It was not the bright idea of a genius to dump radioactive and hazardous chemicals into swamp lands, but this is also only a contributing factor.

Destroying the ozone layer of the atmosphere is also a contributing factor to the global warming effect, but this is also not the sole cause.

The increase in the level of carbon monoxide in the environment might have made irresponsible billionaires ridiculous amounts of money, but this effect works in combination with the previously mentioned factors.

Look at the Earth, metaphorically, as a computer system.

You might be able to do without a single capacitor, and yet the computer might function normally.

If a single component goes out on your system, like a file storage drive or a communications interface, then the computer might still be capable of performing the work assigned to it.

However, if you lose too many sectors of your spinning magnetic media, if too many resistors and capacitors are lost, if too many communications bus lines break, eventually your system will destabilize, and then collapse.

The earth is like that. If we had broken one feature of the Earth, then it would probably heal and be fine.

However, the evil corporations and the monstrous billionaires who run them were not satisfied until they had done major damage to our environment, and to our DNA. Greedy, corrupt, and moralless, they have done their best to destroy everything created by God and Nature.

Ironically, my theory as to another contributing factor has nothing to do with evil corporations and billionaires, human irresponsibility, nor the actions of anyone on this planet.

Data collected by Schpleee Technologies Inc indicates that the primary and inescapable cause of such a sudden rise in the Earth's temperature has to do with a perfectly natural process involving the normal life and behavior of a star, our star, which we call Sol or the Sun.

Chapter 2:
Solar Life

I would have titled this chapter, "Solar Dynamics", however current human technology has so far actually learned very little about solar dynamics, and details of the behavior of a star as it lives and dies.

Over the course of the life of any star, internal nuclear processes consume the material contained within, as power and energy in the form of various kinds of radiation, such as electro-magnetic, are emitted. Eventually, many stars go nova, and the systems of planets orbiting them are destroyed.

In the short term, minor fluctuations of a star's corona, as well as the amount of radiation produced by a star are normal.

Even when a star is not undergoing a major transformation, on a small scale the amount of light and heat emitted by a star can fluctuate, which affects any celestial bodies or objects which orbit the star, or are close enough to receive radiation from it.

As our star ages, the amount of heat, light, and radiation which the Earth receives will ebb and flow, based on the variation in the intensity of the nuclear processes within our sun.

Since life on our planet is completely dependent upon the health of our local star, the condition and output of our sun is of importance.

Many of our astronomical resources are directed towards outer space, and trying to learn as much about the cosmos as possible.

There are comparatively few scientists directing their attention towards that which we see every day - our own sun.

However, as we are now entering a period of increased sun spot activity, and are now seeing greater amounts of radiation bathing our planet, more resources are being dedicated to studying and researching solar activity in our system.

Schpleee Technologies Inc. has been collecting solar data since 2005, with results that indicate an alarming situation manifesting.

Chapter 3:
Solar Resonant Frequency

All heavenly bodies emit a frequency based upon the processes contained within.

From a distance, the Earth also emits a "hum", due to everything happening on the planet, including life and atmospheric turbulence.

We humans are not aware of the "hum" of the Earth, but that is because we are a part of it, and we live within it. Even the noises you make as you live your daily life contribute to the overall pitch of the planet Earth. I call this sound the Resonant Frequency of a heavenly object. Even asteroids and comets each have their own unique characteristic resonant frequency.

Stars, with their massive emissions of power and radiation, and their internal nuclear processes, emit a very powerful resonant frequency.

It has been the resonant frequency data of our sun which has been the focus of the research of Schpleee Technologies Inc.

We at Schpleee have devised a system that collects measurements of the sun's electro magnetic spectrum and analyses it in terms of the sun's resonant frequency.

Of course, it is not possible to hear the sound produced by our sun, because sounds are only perceptible in an atmospheric environment.

However, the digital stream of EM data is converted into an analog signal using a digital to analog converter, and that analog signal can then be played on any amplified transducer.

The effect is that one can "hear" the sound of the hum of the sun, which is both intriguing and somewhat eerie.

Our system sends this analog signal to a sand-vibration table. The sun's resonant frequency then makes geometric patterns in the sand, and a visual analysis of the patterns is made possible.

Pictures of the resonant frequency's geometric pattern are taken over time, and then subjected to computer analysis.

The results which have been obtained so far indicate that the pattern is becoming increasingly complex, and that the hum produced by the sun is going up in frequency.

This indicates that the kinetic energy contained in the sun is increasing, and that the intensity of the nuclear processes in the sun is also currently on the rise.

Data is also being taken by a computer driven telescope which is pointed directly at the sun, and again pictures taken over the last five years have been subjected to computer analysis, which compares the visible portion of the electromagnetic spectrum.

The results of this analysis indicate that the light being emitted by the sun has shifted slightly towards the hot end of the spectrum.

This also coincides with the indication that the sun itself is getting hotter.

The process of Solar Warming has an impact on the entire solar system.

Chapter 4:
Solar System Warming

The effect of Global Warming on the Earth is indeed one of importance to the Human species at this point in history.

This effect however might be caused by Solar Warming, which has an impact on our entire Solar System.

As our local star continues to increase the amount of energy and radiation it is emitting, all heavenly bodies receiving radiation from it will experience a warming effect.

Planets in our solar system which have no electro-magnetic shield like the Earth does will probably suffer an even more pronounced effect due to Solar Warming.

Even the proto-planets farthest away from the sun, such as Pluto, are receiving increased amounts of heat, light, and radiation from the sun.

An increase in the energy being output by a star that is significant enough to have a serious impact on its orbiting planets normally takes millions of years.

However, the analysis of our sun's resonant frequency indicates that another dissonant frequency might be involved which is causing the acceleration of the normal rate of the sun's increase in overall output energy.

I would love to blame the billionaires and their corporations for destroying not only the Earth but the Sun as well, but it is unlikely that the effect of Earth's global warming would have any serious impact on the sun, simply because it is too small in mass compared to the sun to make any difference.

A force with a frequency strong enough to effect an entire star would have to be enormous. It appears to be a measurable effect, but the source of the dissonant frequency is not perceivable by human beings.

One theory is that the source is something that is difficult for our technology to detect, like dark energy.

Another theory is that it is being caused by a force which we cannot see that was created by the explosion of a star, perhaps even in a different galaxy, long ago.

Regardless of what is causing the effect, it is helpful to understand why it is affecting our local star.

For an analogy, take the example of a singer who uses sound to break a glass. It is not the power output by the singer's voice that causes the glass' crystalline structure to shatter. The singer must emit a specific frequency that is at a harmonic that is dissonant to the natural harmonic of the glass to cause it to break.

The harmonic affecting our sun is at a specific frequency that is dissonant to our sun's natural resonant frequency, and it is the combination of these forces that is causing the unnatural accelerated rate of the warming of our sun.

The additional energy output due to this phenomenon is causing all planetary and heavenly bodies in our solar system to experience a warming effect.

Global warming might be posing a threat to human existence, but it is extremely likely that it is impossible for us to do anything to stop it.

Chapter 5:
The End of Planet Earth

Regardless of what you believe is causing the Earth's Global Warming, there is overwhelming data indicating that our planet is heating up at an alarming rate.

Even those not versed in science can see the dramatic lack of a winter at the end of 2011 and the beginning of 2012. On average, this has been the warmest winter in recorded history.

If the abuse of the environment by the corporations is responsible for global warming, then it is still highly unlikely that steps taken to repair the environment will have any visible positive impact in less than fifty years, and it might be too late for humanity by then.

If on the other hand the Solar Warming theory is correct, then the rate that the Earth absorbs solar energy might increase dramatically within the next year.

A sudden increase in the sun's nuclear processes could result in an expansion of the sun's corona, consuming the inner planets of the solar system.

Either way you choose to see it, this might be a good time to invest in sun block, or takes steps to be prepared, anticipating an increase in the temperatures which we are normally accustomed to experiencing.

According to NASA, our sun is experiencing a period of increased solar flare activity, which will continue to increase in intensity for the remainder of 2012.

The Earth is currently facing the possibility of experiencing massive electro-magnetic pulses (EMP) emitted by the sun. If this event occurs, every electrical system on the planet could be destroyed overnight.

Further, although global warming would take years to destroy all multi-cellular life forms, it could quickly lead to world wide disasters such as volcanoes and earthquakes, due to increased tectonic activity, caused by the increase in the temperature of the Earth's core.

Another solar threat to the Earth at this time is that the sun could actually blast a mass of magma towards the Earth, which would certainly wipe out everything except extremophiles, which are bacteria that can survive in extremely hot temperatures.

Not that we can say that any of this is definitely going to happen in the year 2012, but to summarize we are facing the possibility of massive EMP or magma emissions of the sun, increased tectonic forces, and possibly being consumed by the sun's expanding corona.

I am not merely trying to be an alarmist, but if our environment is going to undergo accelerated transformation, I for one want my family and company to be prepared for sudden disaster.

It is advisable for everyone to have a plan as to how they might survive if the culture they are dependent upon vanishes overnight.

Chapter 6:
Surviving Global Disaster

Obviously there is no way of surviving being consumed by the sun, however, it might serve one well to be prepared for lesser forms of global disaster.

If the Earth were to receive a massive electro-magnetic pulse (EMP) from the sun, the results would be devastating to current day human culture. All spinning magnetic media would be erased. This would include most hard drives, many of which are not solid state. There would be no internet, and all cellular networks would be gone.

Worst of all, most of the food eaten in the world these days is made, processed, and transported using computers. Without computers, there would be no fuel to power internal combustion vehicles, and there would be no power for electric cars. Therefore, the food transportation system would be gone.

Most of what an average human does on a daily basis is dependent upon computers. The average American would be lost if suddenly there were no electricity.

Of the Earth's now nearly eight billion human beings, a majority live in cities and urban areas. Most of their food is created and made available to them using computers and electricity.

In the case of a massive EMP, the first effect upon humanity would be starvation.

All of the stores and restaurants would be quickly depleted of edible material.

Maybe it is a blessing that we have so many pets in America, but when people start starving, they are going to start eating pets within a month.

Another month later, most cities will be stripped of food stores and pets, and people are going to have to start eating human beings. Unfortunately meat from a starving human being is very poor nutrition.

If you already are a butcher, you might do well to get a diagram of a human, and develop the techniques involved in processing a human body into steaks and chops.

If you are a cook, it might be an advantage to start developing steak rubs and marinades that taste good with human meat. Here is one now:

Long Pig Marinade

Wine Vinegar	1/3 Cup
Worcestershire Sauce	1/3 Cup
Soy Sauce	2 Tablespoons
Lime Juice	1 Tablespoon
Garlic Juice	¼ Teaspoon
Fresh minced Garlic	2 cloves
Garlic Salt	1 Tablespoon
Chili Powder	1 Teaspoon
Garlic Powder	1 Teaspoon
Ground Cumin	½ Teaspoon
Crushed Onion	¼ Cup
Tabasco Sauce	2 Dashes
Black Pepper	½ Teaspoon

I suggest marinating your human (also known commonly in cannibalistic cultures as "Long Pig") steaks and chops for 48 hours. When you put your marinated steaks on the grill, cook them at a low temperature and baste them with the marinade as they cook. Apply the marinade with a basting brush every time you turn each steak. That will create a nice glaze on the outside of the steaks, just as the marinating process infuses the flavors into the flesh prior to cooking.

On a personal note, you might also discuss with your family who is to be eaten first in the event that you have to eat each other. It could be your family's Order of Human Consumption Plan.

The main problem is that if you live in a city and your life is completely dependent upon retail outlets, you are a goner. However, were I living in a city, I would still build a greenhouse, or if I had no lawn (like in an apartment), I would build indoor horticultural facilities, so that I could grow food.

The people who have the best chance to survive are the ones who can produce as many of their own products and services as possible, like on a homestead farm.

Try to live your life for just one day and attempt to produce everything that you need by yourself. Can you even start a fire without a lighter or matches?

Your basic needs will include food, shelter, and warmth. In the city, the few resources that are on hand will be consumed quickly in the event of a disaster.

It is exactly the ability to *produce* products rather that just *consume* products that will be the key to survival in the even of cultural collapse.

Human beings have been surviving on this planet for thousands of years before modern day electronic technology existed.

Techniques for surviving without electricity were successful; therefore it is possible for us to embrace older technologies and live comfortable lives without video games, cell phones, or the internet.

If you live in the country, you can harvest wood for fuel manually from the forest, and use it to cook the meat that you hunt, the vegetables you grow, and to heat your home. If you own beehives, you have a source of honey for food and wine, and wax for candles. With nothing but a farm, some natural resources, and hard work you can build a home and a life, and supply food and warmth for your family. You and your family would adapt to educate and entertain each other.

While modern day technology is still available, if at all possible, equip your farm with renewable forms of energy, such as windmills and solar panels.

For anyone who can afford it, I strongly recommend building a disaster shelter, equipped with dry goods, greenhouse kits, and mushroom growing facilities. Why mushrooms? They are a source of food that requires almost no light to produce. Stock up on fresh water too, but also build a water distillery in case you need to get it from the river. In the event of disaster, have a plan in place to get everyone and everything that you care about into the shelter in no more than twenty minutes.

No person can produce every single product that they are accustomed to using. I strongly suggest stocking up on non-perishable items that you need, and things that last a long time without electricity, such as canned goods. Also, can your own food. It can remain edible for up to several years if you are lucky. Also, stock up now on hand tools; they can last you a lifetime and require only muscle and physical energy to operate.

I am sure that modern day technologies would be missed, but you would have a means of survival, and in time the technological infrastructure of society could be rebuilt, if humanity manages to survive.

Chapter 7:
Make Your Peace

Well all die someday, but it looks as though perhaps someday soon might all perish together.

If this is the case, then humanity might be preparing to face extinction. In addition to that, many people will try to reconcile their existence in a philosophical or religious light.

For those who believe in a religion, they have one or several Gods to make peace with before facing non-existence.

Atheists and people with other philosophies might contemplate their place in nature, and the impact that humanity has had on this world. In retrospect, the existence of humanity on the Earth was no "gift from God". The billionaires and their corporations have attacked and destroyed virtually every single creation of God and Nature.

The relationship of humanity to each other also seems to be of importance.

It is said that humans are social creatures. This is not always true, as throughout the ages some people have chosen to live as hermits.

However, a majority of humans have lived most of their lives socially. It is the association with the people whom we spend time with every day that forms the bonds in our hearts, and make our existence feel meaningful to us.

I suggest going out of your way to make sure that the people you care about the most are near your geographically in your daily life. If disaster does strike, there is an excellent chance that you will get little warning or none at all.

Like any mortal being, I know that I will die some day, but when that day comes, I don't want to be alone, nor do I want those whom I love to die without me. If we are all going out, I plan to be with those I love, and being with them will make me feel safe.

Chapter 8:
Summary and Conclusion

Using a makeshift computer driven "sonar telescope", Schpleee Technologies Inc has been collecting data pertaining to the resonant frequency of the Sun, the local star in our solar system.

The signal is converted from digital data into an analog signal that is used to supply signal to a sand-vibration table, and pictures of the resulting geometric pattern are computer analyzed to track changes over time.

The results have shown that the sun's resonant frequency is going up, as is its kinetic energy.

A computer analysis of the sun's visible spectrum over time has also shown that the sun is getting hotter, which is causing a heating of all planetary bodies system wide.

The phenomenon of Solar Warming is causing an increased temperature in the Earth's core, which is causing an increase in tectonic activity world wide, and an increase in related disasters such as volcanic activity and earthquakes.

If a solar event totally destroys the Earth, then our fate is completely out of our hands.

However, even if this does not happen it is likely that we are yet to face disaster anyway, and it is better to be prepared than to find oneself suddenly helpless without retail stores.

The key is to produce rather than to consume. Stock up on things that you cannot produce prior to a possible disaster.

Create a homestead in which you and your family can create what you need to survive. Make sure that your family has an emergency disaster plan in which you can provide food, warmth, and shelter for your family without electricity and in a sustainable manner.

Make your peace with Nature or whatever God or Gods you believe in, and keep your family and loved ones nearby.

In the event of disaster, you can face it with those whom you love.